Using Derivative Spectrophotometry for Determination of Some Drugs

Sahar Fadhel

Using Derivative Spectrophotometry for Determination of Some Drugs

LAP LAMBERT Academic Publishing

Imprint

Any brand names and product names mentioned in this book are subject to trademark, brand or patent protection and are trademarks or registered trademarks of their respective holders. The use of brand names, product names, common names, trade names, product descriptions etc. even without a particular marking in this work is in no way to be construed to mean that such names may be regarded as unrestricted in respect of trademark and brand protection legislation and could thus be used by anyone.

Cover image: www.ingimage.com

Publisher:
LAP LAMBERT Academic Publishing
is a trademark of
International Book Market Service Ltd., member of OmniScriptum Publishing Group
17 Meldrum Street, Beau Bassin 71504, Mauritius

ISBN: 978-613-9-83025-1

Copyright © Sahar Fadhel
Copyright © 2018 International Book Market Service Ltd., member of OmniScriptum Publishing Group

1. Simultaneous determination of olanzapine, trifluoperazine hydrochloride and carbamazepine via derivative spectrophotometry

1.1 Introduction

The term derivative mode is used to describe spectra that are obtained by plotting the higher order derivative of absorbance according to their wavelength Figure (1.1), the first and second order derivative spectra were used initially, but with evolution of modern electronics, most of the commercial spectrophotometers were fitted with derivative modules to resolve signals with higher-order differentiation (n > 2).[1,2]

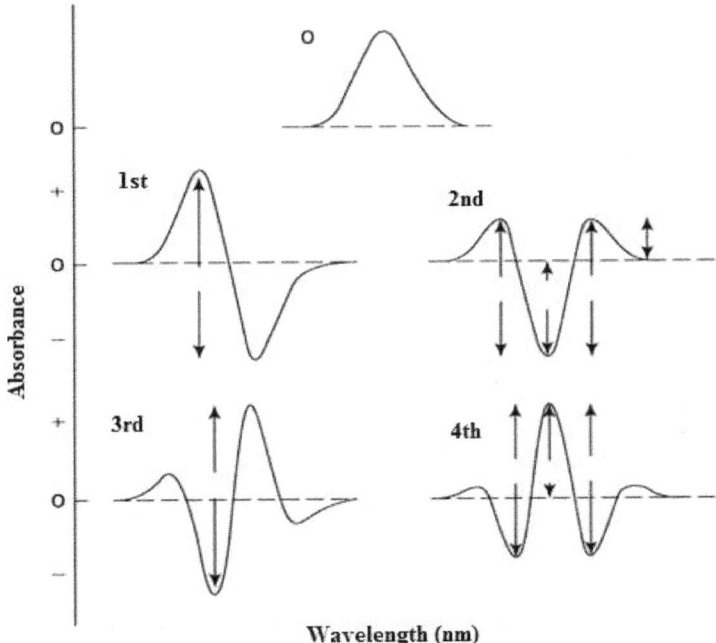

Figure (1.1): The shape of 0, 1st, 2nd, 3rd and 4th order derivative spectrum.

Derivative spectrophotometry is used for qualitative and quantitative analysis, and for eliminating the effect of baseline shifts and baseline tilts. It consists of calculating and plotting one of the mathematical derivatives of a spectral curve; this feature leads to narrowing bands and as a consequence to separate the overlapped peaks. Derivative spectrophotometry is now a reasonably prized standard feature of modern micro-computerized UV spectrophotometry.[3-5] Derivatization of spectral data was first introduced in the 1953 [3] but in 70s of the twentieth century, a fast development of this technique started, when spectrophotometers were controlled by computers. In 80s of the last century a peak of spectrophotometers popularity had occurred. In the present time, only additional technique has been used as a build-in function in software of modern spectrophotometers.[4]

For the determination of two or more active compounds simultaneously in the same mixture without a separation step, many spectrophotometric methods were used. Derivative spectra could be obtained by optical, electronic, or mathematical methods. Optical and electronic techniques were used on early UV-Vis spectrophotometers but have largely been superseded by mathematical techniques.[6,7]

This technique is based on measuring changes in intensity or absorbance, manually or automatically. The base of this approach is that the wavelength scan rate, $(d\lambda/dt)$ is constant, and then the derivative of the intensity with respect to the wavelength; $(dI/d\lambda)$ is proportional to the derivative of the intensity with respect to the time, (dI/dt), which is measured by means of its electronic differentiation:[8,9]

$$(dI/d\lambda) = (dI/dt)/(d\lambda/dt)$$

Derivative spectra possess the same relation to the concentration of the absorbed species as that in the original normal (zero order) spectrum i.e. Beer's law is obeyed. In addition, concentration measurement of an analyte in the presence of interference or of two or more analytes in a mixture can sometimes be made more easily or more accurately using derivative methods. Derivative spectrum of n-component mixture is a sum of derivative spectra of individual components. [4, 10]

Derivative spectrophotometry is an analytical technique of major usefulness for extracting the qualitative and quantitative information from spectra composed of unresolved bands by using the first or higher derivatives of absorbance with respect to wavelength. This technique offers various advantages over the conventional absorbance methods, like recognition of the sharp spectral features over the large bands and increasing the resolution of overlapping spectra and allowance the assay of certain analytes from complex mixtures or matrices via mathematical interpretation of the absorption signal. [9,11]

Derivative spectrophotometry has a rule in different aspects, like organic compounds, inorganic analysis, analysis of biological compounds, pharmaceutical analysis, food analysis and environmental analysis. [12]

1.2 Estimation of some drugs in dosage compounds and human fluids by derivative mode

Initially derivative spectrophotometric method was utilized for simultaneous determination of metal ions mixture and its applications have been extend to other analytical fields; among them is the pharmaceutical compounds. [13]

Derivative spectrophotometric method is one of the analytical techniques that have a great utility for resolving drug mixtures with overlapping spectra. Moreover, derivative methods have been applied successfully to the determination of drugs in the presence of their degradation products. Also derivative techniques have proven to be very useful in the resolution of binary and ternary mixtures.[9] Table (1.1) shows the analytical characteristics of these methods for the simultaneous determination of some pharmaceutical compounds.

Table (1.1): Analytical study of derivative methods for simultaneous determination of some drugs in pharmaceutical products and human fluids.

Pharmaceutical products	λ_{max} (nm)	Derivative order and application remark	Ref. No.
Metronidazole and nystatin	290 and 261	1st, in their mixture	12
Metformin HCl and Pioglitazone HCl	279.5 and 247.5	2nd, in a bilayer tablet formulation	13
Bifonazole	223	1st, Cream, tablets	14
Ezetimibe and Atorvastatin calcium	224.6, 223.8, 233.0 and 238.6	1st, 2nd, tablets	15
Moxifloxacin HCl and Cefixime trihydrate	287 and 317.9	1st, in their mixture and in presence of tablet excipients	16
Diazepam	313	4th, Human plasma	17
Irbesartan	244 and 230	3rd, 4th, bulk drug and tablets	18
Rifampicin and Piperine	238 and 272	1st, Capsule dosage form	19
Paracetamol, Ibuprofen and Caffeine	242.4, 271.2 and 302.4	1st, 3rd, mixture of tablet	20
17-Beta-Estradiol and Drospirenone	208 and 282	1st, Combined Dosage Form and tablet	21
Paracetamol and Tramadol	249 and 271	1st, 2nd, mixture of tablets	22
Gatifloxacin and Prednisolone acetate	243 and 268	2nd, Eye drops	23

| Acetaminophen, Diphenhydramine and Pseudoephedrine | 281.5, 226.0 and 218.0 | 1st, 2nd, 4th, mixture of tablets | 24 |
| Paracetamol and Ibuprofen | 274.8 and 234.4 | 1st, 2nd, Combined tablets | 25 |

1.3 Experimental

1.3.1 Instruments

The instruments used during in this study are listed in table (1.2).

Table (1.2): The instruments used in throughout the study.

No.	Instrument	Source and model
1.	UV-Visible double beam spectrophotometer with 1ml quartz cell	Shimadzu -1800, Japan
2.	UV-Visible double beam spectrophotometer with 1ml quartz cell	PG instrument –T80, U.K
3.	Analytical balance (±0.0001 g)	Sartorious BL 210 S Scientific balance, Gottingen - Germany.
4.	Thermostatic Shaker water bath	LabTech Moldel LSB-015S

1.3.2 Chemical compounds

The chemical compounds, which were used in this work and their suppliers, are listed in table (1.3)

Table (1.3): Chemical compounds and their supplier.

No.	Materials	Purity or assay	Supplier
1.	Ethanol	99.5 %w/w	Sigma-Aldrich
2.	Methanol	99.9 %w/w	Scharlau
3.	Glucose	Pure powder	BDH
4.	Lactose	Pure powder	BDH
5.	Sucrose	Pure powder	BDH
6.	Starch	Pure powder	BDH
7.	Magnesium Stearate	99.9% w/w	BDH

1.3.3 The drugs and pharmaceutical preparations

Pharmaceutical grade olanzapine, trifluoperazine hydrochloride and carbamazepine powders received in pure form (99.99%) were provided as a gift from the State Company for Drug Industries and Medical Appliances Samarra-Iraq (S.D.I). The pharmaceuticals formulations that were used in this work are summarized in table (1.4) with their corresponding manufacturers.

Table (1.4): Pharmaceuticals formulations and their manufacturer companies.

No.	Pharmaceutical formulation	Manufacturer companies
1.	Zyprexa® -5 mg /Tablet	Lilly – Spain
2.	Zyprexa® -10 mg /Tablet	Lilly – Spain
3.	Olan® – 5 mg / Tablet	Micro – India
4.	Iralzin® – 5mg /Tablet	S.D.I. – Iraq
5.	Espazine® – 5mg / Tablet	GSK – England
6.	Stellasil® – 1mg / Tablet	Kahira pharma – Egypt
7.	Carbasam® – 200 mg / Tablet	S.D.I. - Iraq
8.	Tegretol® – 200 mg / Tablet	Novartis - Switzerland

1.3.4 Preparation of solutions

- Glucose [10000 µg.mL^{-1}]: prepared by dissolving 0.1 g of glucose in a suitable volume of distilled water and the volume was made up to the mark in 10 mL volumetric flask.

- Lactose [10000 µg.mL^{-1}]: prepared by dissolving 0.1 g of lactose in a suitable volume of distilled water and the volume was made up to the mark in 10 mL volumetric flask.

- Sucrose [10000 µg.mL^{-1}]: prepared by dissolving 0.1 g of sucrose in a suitable volume of distilled water and the volume was made up to the mark in 10 mL volumetric flask.

- Magnesium stearate [10000 µg.mL^{-1}]: prepared by dissolving 0.1 g of Mg(C$_{18}$H$_{35}$O$_2$)$_2$ in a suitable volume of distilled water and the volume was made up to the mark in 10 mL volumetric flask.
- Starch [10000 µg.mL^{-1}]: 0.1000g of starch was triturated with a small amount of cold water into a thin paste, then 10 mL of boiling water were added to the paste. The mixture was boiled for about 5 minutes until a clear solution is obtained. This solution should be freshly prepared as required.[26]

1.3.5 Preparation of standard drugs solutions

1.3.5.1 Olanzapine stock solution (1000 µg.mL^{-1})

The stock solution of (OLN.) was prepared by dissolving an accurately weighed 0.0500g of pure drug in a suitable volume of ethanol and the solution was made up to the mark in 50mL volumetric flask with ethanol. The stock solution was protected from light and stored at 5°C.

Olanzapine working solution (100µg.mL^{-1}): prepared by diluting 10 mL of the stock solution to the mark with ethanol in a 100mL volumetric flask.

1.3.5.2 Trifluoperazine hydrochloride stock solution (1000 µg.mL^{-1})

The stock solution of (TFZ.) was prepared by dissolving an accurately weighed 0.0500g of pure drug in a suitable volume of ethanol and the solution was made up to the mark in 50mL volumetric flask with ethanol. The stock solution was protected from light and stored at 5°C.

Trifluoperazine hydrochloride working solution (100 µg.mL^{-1}): prepared by diluting 10 mL of the stock solution to the mark with ethanol in a 100 mL volumetric flask.

1.3.5.3 Carbamazepine stock solution (1000 µg.mL^{-1})

The stock solutions of (CRN.) was prepared by dissolving accurate weighed 0.1000 g of pure drug in 10 ml of ethanol and the volume was made up to the mark in volumetric flask 100 mL with ethanol. The stock solutions were stored at 5°C and were protected from light.

Carbamazepine working solution (100 µg.mL^{-1}), was prepared by dilution of 10 mL of the stock solution, the volume is completed with ethanol to 100mL in a volumetric flask.

1.3.6 Solutions for the analysis of drugs in pharmaceutical preparations
1.3.6.1 Olanzapine

The contents of 7 tablets of Zyprexa (5mg), Zyprexa (10mg), and 10 tablets of Oln-5 were accurately and separately weighed, ground into fine powder and mixed well, then an average weight was calculated. An amount of the powder equivalent to 0.5928 g, 0.2972 g and 0.5848 g (containing 0.0200g of olanzapine drug) of Zyprexa-5mg, Zyprexa-10mg and Oln-5mg respectively was accurately and separately weighed, dissolved in a suitable volume of ethanol and stirred for 10min to ensure complete dissolution of the drug. The solution was transferred into 20mL volumetric flask and diluted to the mark with ethanol to get 1000 µg.mL^{-1} of (OLN.). The solution was then filtered by using Whatman filter paper No.41 to avoid any suspended or un-dissolved material before use.

Working solution 100µg.mL^{-1} was freshly prepared by dilution with ethanol and analyzed by the recommended procedure.

1.3.6.2 Trifluoperazine hydrochloride

The content of 10 tablets for Iralzin-5mg, Espazine-5mg and 20 tablets for Stellasil-1mg was accurately and separately weighed, ground into fine powder and mixed well then the average weight was calculated. An amount of the powder equivalent to 0.5234g, 0.5194g and 1.0950g (containing 0.0200g of trifluoperazine hydrochloride drug) for Iralzin-5mg, Espazine-5mg and Stellasil-1mg respectively was accurately and separately weighed, dissolved in a suitable volume of ethanol and stirred for 10min to ensure complete dissolution of the drug, then transferred into 20mL volumetric flask and diluted to the mark with ethanol to get 1000µg. mL^{-1} of (TFZ.). The solution was filtered by using Whatman filter paper No.41 to avoid any suspended or un-dissolved material before use.

Working solution 100µg.mL^{-1} was freshly prepared by dilution with ethanol and analyzed by the recommended procedure.

1.3.6.3 Carbamazepine

The content of 10 tablets of Carbasam (200 mg) and Tegretol (200 mg) were accurately and individually weighed, grinded into fine powder then mixed well and an average weight was calculated for each drug. An amount of the powder equivalent to 0.0842 g and 0.0839 g (containing 0.0500 g of the drug carbamazepine) of Carbasam-200 mg and Tegretol-200 mg respectively was accurately and separately weighed, dissolved in 10 ml ethanol and stirred for 10 min to ensure complete dissolution of the drug, then transferred into 50 mL volumetric flask and diluted to the mark with ethanol to get 1000 µg.mL^{-1} (CRN.). The solution was filtered by using Whatman filter paper No.41 to avoid any suspended or un-dissolved material before use.

Working solution (250 µg.mL^{-1}) was freshly prepared by dilution in the ethanol and analyzed by the recommended procedure.

1.3.7 Recommended procedures

1.3.7.1 Assay procedure for the determination of olanzapine, trifluoperazine hydrochloride and carbamazepine

1mL aliquots, of olanzapine standard solution containing (10- 70) µg or trifluoperazine hydrochloride standard solution containing (10- 65) µg or carbamazepine standard solution containing (10- 70) µg, were transferred into a series of 5 mL volumetric flask, and diluted to the mark with ethanol. The spectrum for each solution was recorded against ethanol. Zero order spectrums were then manipulated for each drug to get its first derivative (D1) and second derivative (D2).

1.3.7.2 Assay procedure for the determination of the drugs mixtures

- **Olanzapine**

1 mL aliquots, of (OLN.) standard solution containing (10- 70) µg were transferred into a series of 5 mL volumetric flask containing 1 mL of (10, 30 and 50) µg of (TFZ.) solution and 1 mL of (10, 30 and 50) µg of (CRN.), the mixture was then diluted to the mark with ethanol. The spectrum for each solution was recorded against ethanol. The recorded spectra were then manipulated to get (D1) and (D2).

- **Trifluoperazine hydrochloride**

1 mL aliquots, of (TFZ.) standard solution containing (10- 65) µg were transferred into a series of 5 mL volumetric flask containing 1 mL of (10, 30 and 50) µg of (OLN.) solution and 1 mL of (10, 30 and 50) µg of (CRN.), the mixture was then diluted to the mark with ethanol. The spectrum for each solution was recorded against ethanol. The recorded spectra were then manipulated to get (D1) and (D2).

- **Carbamazepine**

1 mL aliquots, of (CRN.) standard solution containing (10- 70) µg were transferred into a series of 5 mL volumetric flask containing 1 mL of (10, 30 and 50) µg of (OLN.) solution and 1 mL of (10, 30 and 50) µg of (TFZ.), the mixture was then diluted to the mark with ethanol. The spectrum for each solution was recorded against ethanol. The recorded spectra were then manipulated to get (D1) and (D2).

1.4 Results and discussion

1.4.1 Absorption spectra at zero order mode

The absorption spectra of (OLN.), (TFZ.) and (CRN.) and their mixture were recorded against ethanol as a blank. The absorption spectrum of (OLN.) has maximum wavelength of absorption at (272 and 226) nm, (TFZ.) has maximum wavelength of absorption at (310, 260 and 214) nm, and the absorption spectrum of (CRN.) which appears absorption maxima at (284, 237 and 216) nm in addition to the absorption spectrum of mixture of three drugs which show a maximum wavelength of absorption at (261, 238 and 220) nm which is related to the absorption maxima of the three compounds. Figure (1.2) displays the absorption spectrum of (OLN.), (TFZ.), (CRN.) and drugs mixture.

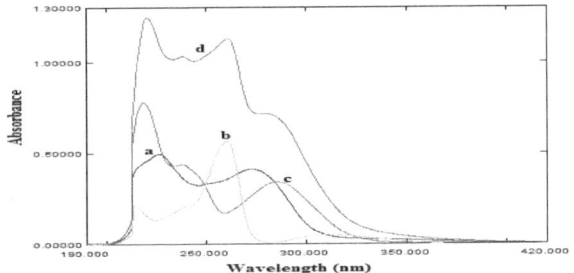

Figure (1.2): Absorption spectra of **(a)** 6μg.mL^{-1} (OLN.), **(b)** 4μg.mL^{-1} (TFZ.) and **(c)** 4μg.mL^{-1} (CRN.), and **(d)** a mixture of 6 μg.mL^{-1} (OLN.), 4μg.mL^{-1} (TFZ.) and 4μg.mL^{-1} (CRN.).

1.4.2 First and second derivative modes

Large overlap of the spectra of (OLN.), (TFZ.), and (CRN.) are obvious. Therefore, the direct determination of the drugs using zero order absorption measurements, when they are present in the same solution is very difficult using conventional methods. Derivative spectrophotometric technique, as mentioned before, is of a particular utility in the determination of the concentration of single component in such mixtures, with large spectral overlapping. For this reason, derivative spectrophotometric methods have been applied. To select the derivative order, the first and second derivative spectra of (OLN.), (TFZ.) and (CRN.) were recorded. The investigation reveals that first and second order spectra were simple and gave results of highest accuracy and detection limits. The first and second order derivative spectra of (OLN.), (TFZ.), (CRN.) and their mixture are shown in Figures (1.3) and (1.4) respectively.

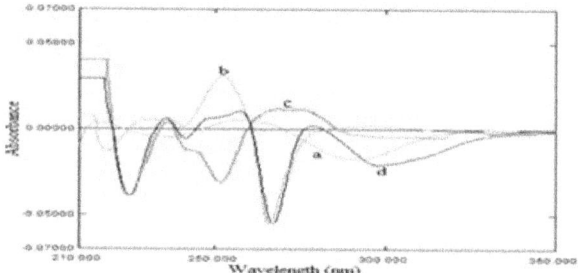

Figure (1.3): First derivative spectra of **(a)** 6 µg.mL^{-1} (OLN.), **(b)** 4 µg.mL^{-1} (TFZ.), **(c)** 4 µg.mL^{-1} (CRN.), and **(d)** a mixture of 6 µg.mL^{-1} (OLN.), 4 µg.mL^{-1} (TFZ.) and 4 µg.mL^{-1} (CRN.).

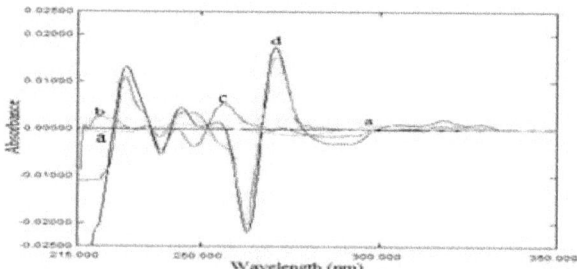

Figure (1.4): Second derivative spectra of **(a)** 6 µg.mL^{-1} (OLN.), **(b)** 4 µg.mL^{-1} (TFZ.), **(c)** 4 µg.mL^{-} (CRN.) and **(d)** a mixture of 6 µg.mL^{-1} (OLN.), 4 µg.mL^{-1} (TFZ.) and 4 µg.mL^{-1} (CRN.).

1.4.3 Calibration curves for olanzapine

Four graphical techniques to determine the values of derivative spectra namely, peak to baseline (peak height), peak to peak, area under the peak and zero crossing, have been used via UV- spectrophotometric method for quantitative analyses of olanzapine individually and in its mixture with trifluoperazine hydrochloride and carbamazepine.

In the baseline to peak technique, the measurement was carried out from a maximum to the zero line or from a minimum to the zero line. In the peak to peak technique, the determination was carried out by measuring the amplitude from a maximum to a minimum of the curve. In area under peak technique, the area was calculated. In zero crossing technique, measurement of the absolute value of the total derivative spectrum at an abscissa value corresponding to the zero-crossing wavelength of the derivative spectra of individual components, which should only be a function of the concentration of other component.

Figures (1.5) to (1.8) show first order spectra for sets of solutions containing various amounts (2-14 µg.mL^{-1}) of (OLN.) in the presence of different amounts (0, 2, 6 and 10) µg.mL^{-1} of (TFZ.) and (CRN.). In the first derivative technique, the results indicate that when the concentration of (TFZ.) and (CRN.) are kept constant and the concentration of (OLN.) varied, the peak amplitudes measured as peak to baseline at 291.47 nm were found to be in proportion with olanzapine concentration. Therefore, the calibration plots were constructed for the assay of (OLN.) in the presence of (TFZ.) and (CRN.) at the mentioned wavelengths by plotting the measured values (as signals) against the concentration of (OLN.). Figures (1.9) to (1.12).

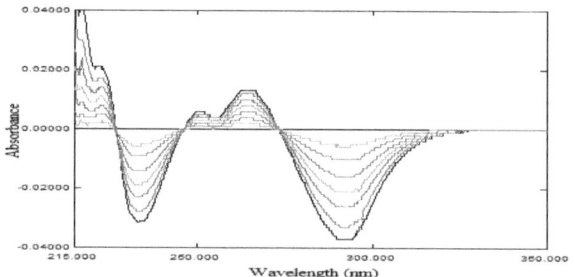

Figure (1.5): First derivative spectra of solutions containing (2-14) µg.mL^{-1} olanzapine.

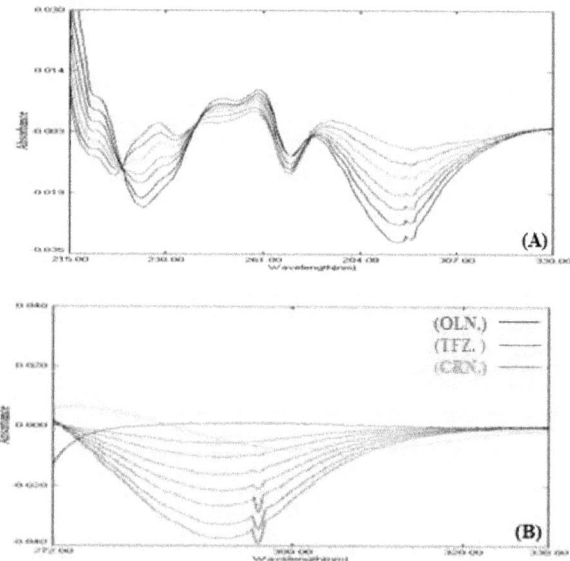

Figure (1.6): First derivative spectra of mixtures contain (2-14) µg.mL^{-1} olanzapine in the presence of 2 µg.mL^{-1} trifluoperazine hydrochloride and carbamazepine (A) at (215 to 330)nm, (B) at (272 to 330) nm.

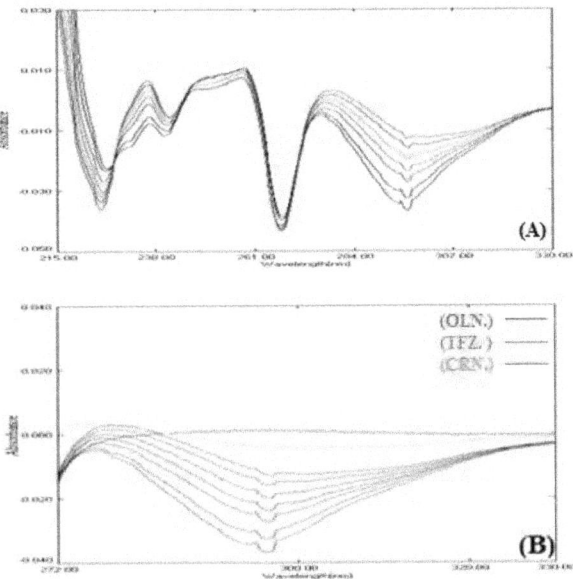

Figure (1.7): First derivative spectra of mixtures contain (2-14) µg.mL^{-1} olanzapine in the presence of 6µg.mL^{-1} trifluoperazine hydrochloride and carbamazepine (A) at (215 to 330) nm, (B) at (272 to 330) nm.

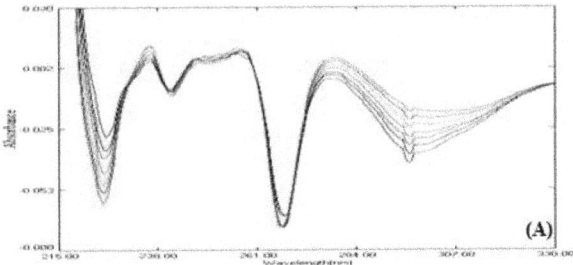

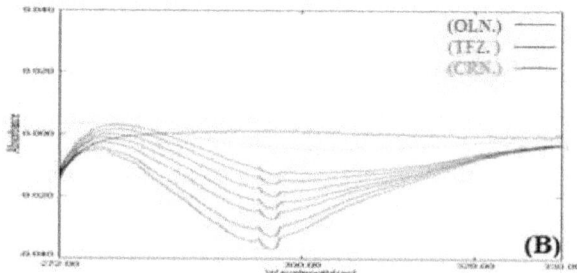

Figure (1.8): First derivative spectra of mixtures contain (2-14) µg.mL^{-1} olanzapine in the presence of 10µg.mL^{-1} trifluoperazine hydrochloride and carbamazepine (A) at (215 to 330) nm, (B) at (272 to 330) nm.

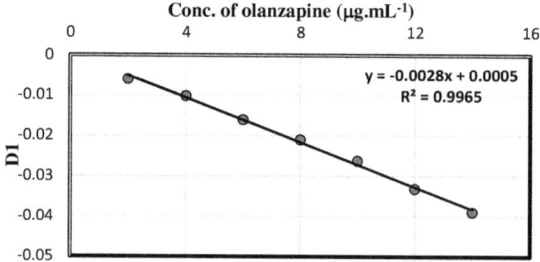

Figure (1.9): Calibration curve obtained via first derivative spectra of (2-14) µg.mL^{-1} olanzapine for peak-to-baseline at 291.47 nm.

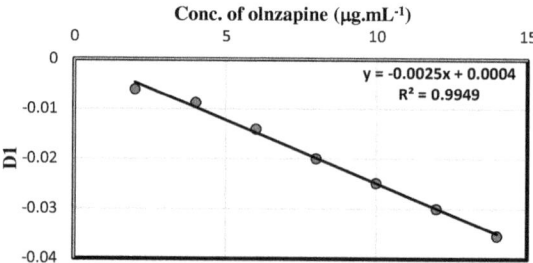

Figure (1.10): Calibration curve obtained via first derivative spectra of Mixtures contain (2-14) µg.mL^{-1} olanzapine in the Presence of 2 µg.mL^{-1} trifluoperazine hydrochloride and carbamazepine for peak-to-baseline at 291.47 nm.

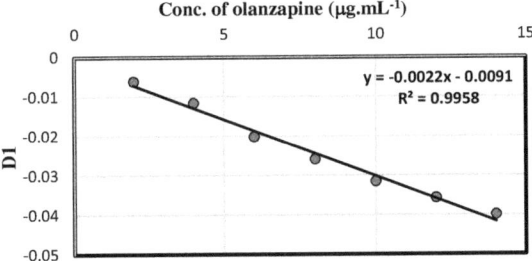

Figure (1.11): Calibration curve obtained via first derivative spectra of mixtures contain (2-14) µg.mL^{-1} olanzapine in the presence of 6 µg.mL^{-1} trifluoperazine hydrochloride and carbamazepine for peak to baseline at 291.47 nm.

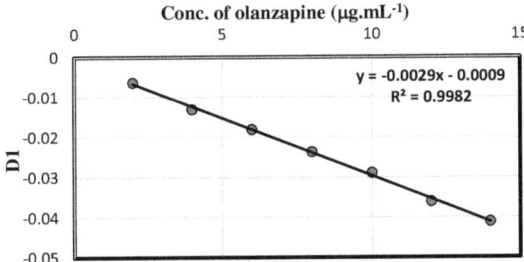

Figure (1.12): Calibration curves obtained via first derivative spectra of Mixtures contain (2-14) µg.mL^{-1} olanzapine in the Presence of 10 µg.mL^{-1} trifluoperazine hydrochloride and carbamazepine for peak to baseline at 291.47 nm.

Moreover, the second derivative spectra of the same sets of solution containing (2-14) µg.mL^{-1} (OLN.) with different spiked concentrations of (TFZ.) and (CRN.) (0, 2, 6 and 10) µg.mL^{-1} were also recorded and attempts were made to utilize them for finding the concentrations of the drug. Figures (1.13)-(1.16) show the second derivative spectra for different concentrations of (OLN.) and for its mixtures with (TFZ.) and (CRN.).

The delicate inspection of the second derivative spectra obtained for (OLN.) and for its mixtures with (TFZ.) and (CRN.) show that height to baseline at zero cross at 300.00 nm measurements could be used to quantify the exact concentration of (OLN.) in the presence of (TFZ.) and (CRN.).

Measurements of height to baseline at zero cross at accurately selected regions in the recorded spectra, were carried out. Calibration plots were constructed when the measured values were plotted against the concentration of olanzapine as shown in Figures (1.17)-(1.20).

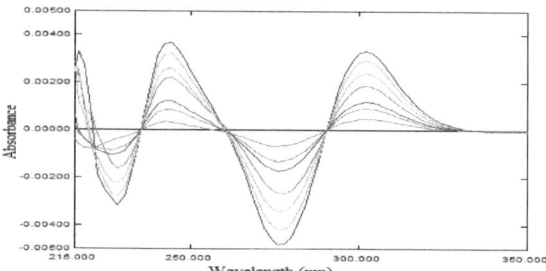

Figure (1.13): Second derivative spectra of solutions containing (2-14) µg.mL^{-1} olanzapine.

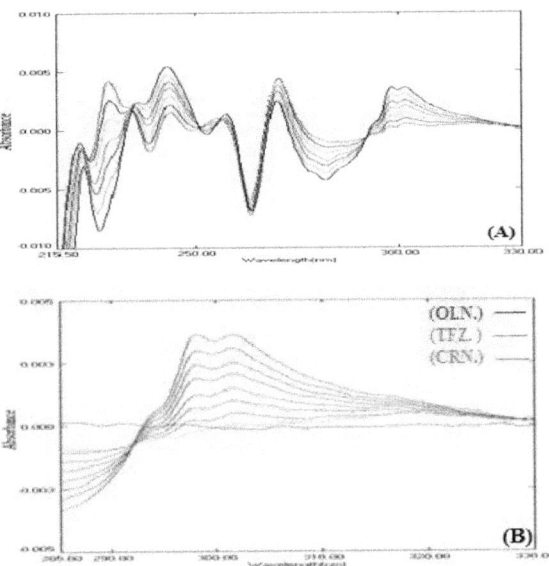

Figure (1.14): Second derivative spectra of mixtures contain (2-14) µg.mL^{-1} olanzapine in the presence of 2 µg.mL^{-1} trifluoperazine hydrochloride and carbamazepine at (A) (215.5-330)nm and (B) (285-330)nm.

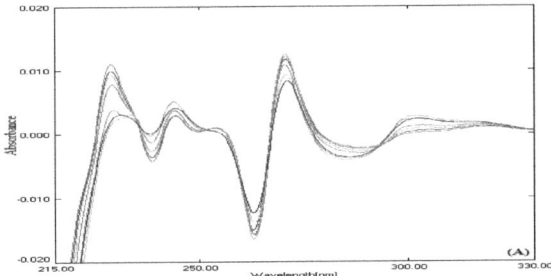

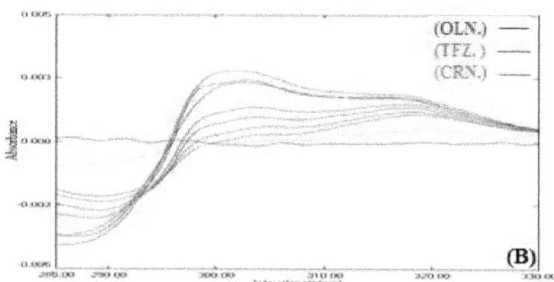

Figure (1.15): Second derivative spectra of mixtures contain (2-14) µg.mL^{-1} olanzapine in the presence of 6 µg.mL^{-1} trifluoperazine hydrochloride and carbamazepine at (A) (215.5-330)nm and (B) (285-330)nm.

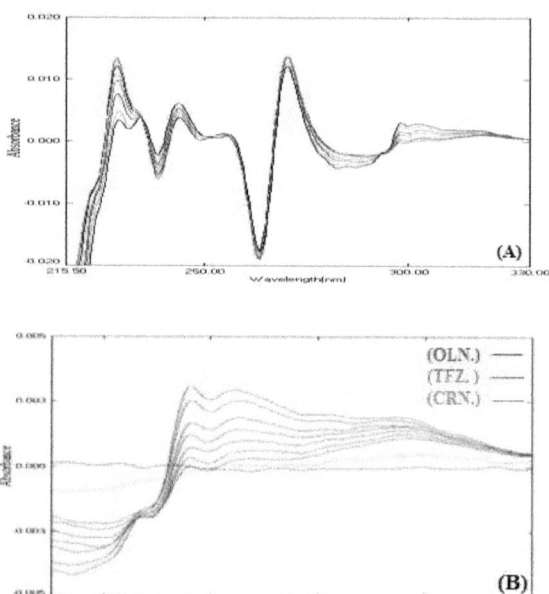

Figure (1.16): Second derivative spectra of mixtures contain (2-14) µg.mL^{-1} olanzapine in the presence of 10 µg.mL^{-1} trifluoperazine hydrochloride and carbamazepine at (A) (215.5-330)nm and (B) (285-330)nm.

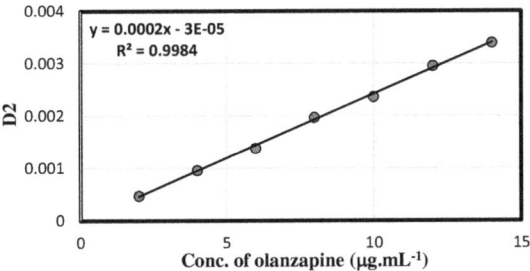

Figure (1.17): Calibration curve via second derivative spectra of (2-14) µg.mL^{-1} olanzapine for height to baseline at zero cross at 300.00 nm.

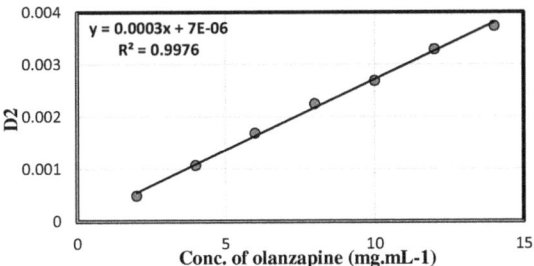

Figure (1.18): Calibration curve via second derivative spectra of mixtures contain (2-14) µg.mL^{-1} olanzapine in the presence of 2 µg.mL^{-1} trifluoperazine hydrochloride and carbamazepine for height to baseline at zero cross at 300.00 nm.

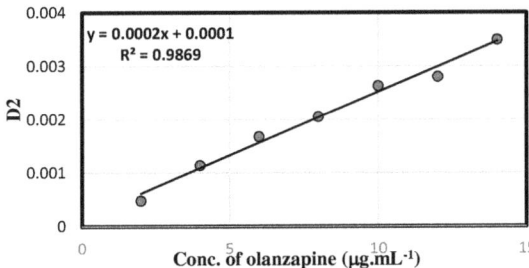

Figure (1.19): Calibration curve via second derivative spectra of mixtures containing (2-14) µg.mL^{-1} olanzapine in the presence of 6 µg.mL^{-1} trifluoperazine hydrochloride and carbamazepine for height to baseline at zero cross at 300.00 nm.

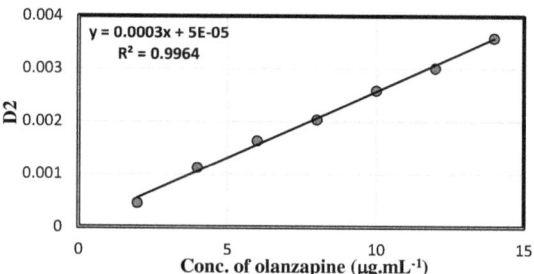

Figure (1.20): Calibration curve via second derivative spectra of mixtures contain (2-14) µg.mL^{-1} olanzapine in the presence of 10 µg.mL^{-1} trifluoperazine hydrochloride and carbamazepine for height to baseline at zero cross at 300.00 nm.

1.4.4 Calibration curves for trifluoperazine hydrochloride

Under the mentioned optimum conditions for D1 and D2 modes in selected ranges of wavelengths, calibration curves were obtained for the assay of (TFZ.) via UV-spectrophotometric method. The recorded spectra using D1 mode for a set of solutions containing (2-13) µg.mL^{-1} (TFZ.) in the presence of different concentrations of (OLN.) and (CRN.) (0, 2, 6 and 10) µg.mL^{-1} are shown in Figures (1.21)-(1.24). The procedure showed good results over the studied range of concentration depending on peak to baseline at 266.95 nm measurements could be used to quantify the exact concentration of (TFZ.) in the presence of (OLN.) and (CRN.), Figures (1.25)-(1.28) show plotted calibration curves.

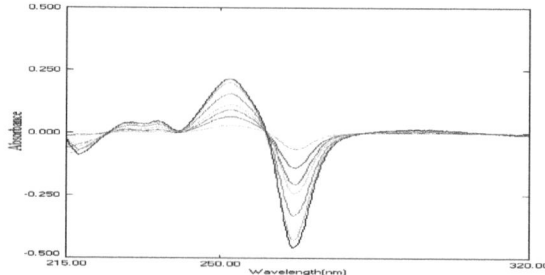

Figure (1.21): First derivative spectra of solutions containing (2-13)µg.mL^{-1} trifluoperazine hydrochloride.

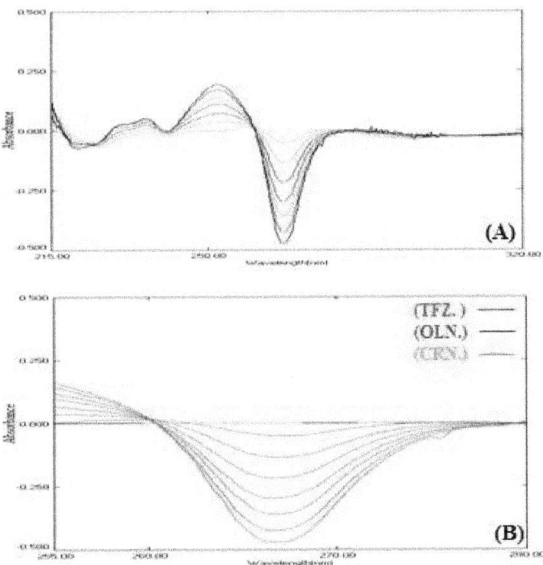

Figure (1.22): First derivative spectra of mixtures contain (2-13) µg.mL^{-1} trifluoperazine hydrochloride in the presence of 2 µg.mL^{-1} olanzapine and carbamazepine (A) at (215- 320) nm and (B) at (255- 280)nm.

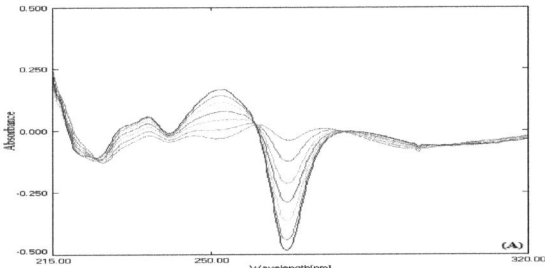

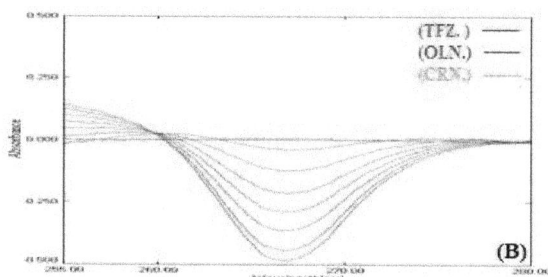

Figure (1.23): First derivative spectra of mixtures contain (2-13) μg.mL^{-1} trifluoperazine hydrochloride in the presence of 6 μg.mL^{-1} olanzapine and carbamazepine (A) at (215- 320) nm and (B) at (255- 280)nm.

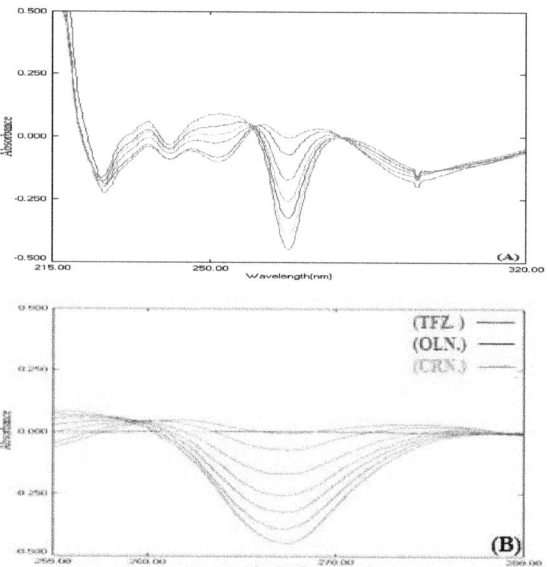

Figure (1.24): First derivative spectra of mixtures contain (2-13) μg.mL^{-1} trifluoperazine hydrochloride in the presence of 10μg.mL^{-1} olanzapine and carbamazepine (A) at (215- 320) nm and (B) at (255- 280)nm.

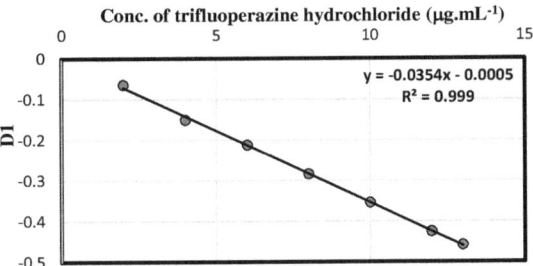

Figure (1.25): Calibration curve obtained via first derivative spectra of (2-13) µg.mL^{-1} trifluoperazine hydrochloride for peak to baseline at 266.95 nm.

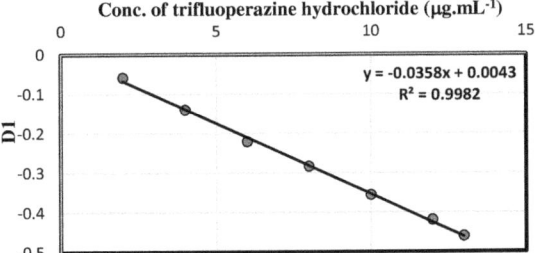

Figure (1.26): Calibration curve obtained via first derivative spectra of mixtures contain (2-13) µg.mL^{-1} trifluoperazine hydrochloride in the presence of 2 µg.mL^{-1} olanzapine and carbamazepine for peak to baseline at 266.95 nm.

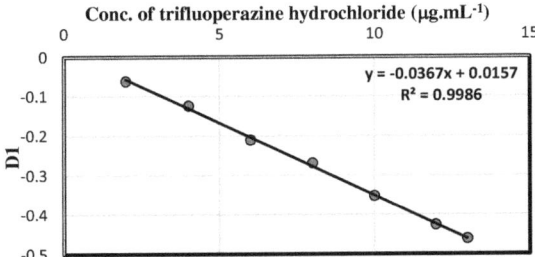

Figure (1.27): Calibration curve obtained via first derivative spectra of mixtures containing (2-13) µg.mL^{-1} trifluoperazine hydrochloride in the presence of 6 µg.mL^{-1} olanzapine and carbamazepine for peak-to-baseline at 266.95 nm.

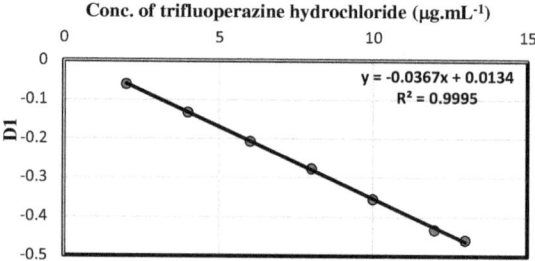

Figure (1.28): Calibration curve obtained via first derivative spectra of mixtures contain (2-13) µg.mL^{-1} trifluoperazine hydrochloride in the presence of 10 µg.mL^{-1} olanzapine and carbamazepine for peak-to-baseline at 266.95 nm.

The second derivative spectra obtained for the mixtures of (TFZ.), (OLN.) and (CRN.) show that peak to baseline measurements at specified wavelength 263.47nm and 270.25nm could be used to quantify the exact concentration of (TFZ.) in the presence of (OLN.) and (CRN.). Figures (1.29) - (1.32) show the calculated second derivative spectra of different mixtures of the cited drugs while, Figures (1.33)-(1.36) represent the calibration graphs for the same mixtures.

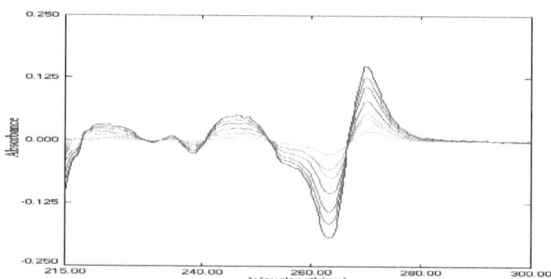

Figure (1.29): Second derivative spectra of solutions containing (2-13) µg.mL^{-1} trifluoperazine hydrochloride.

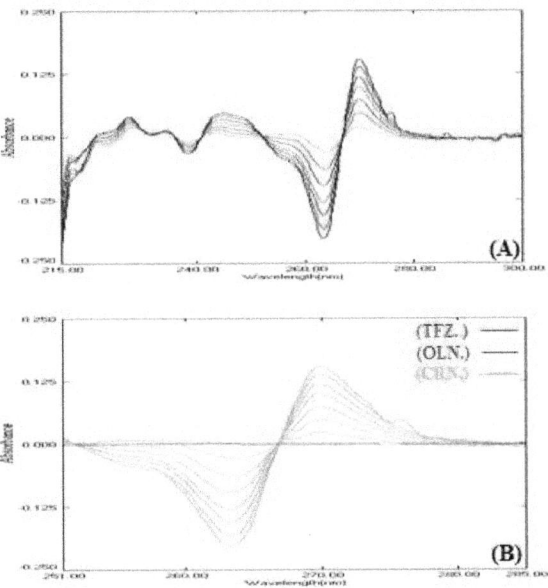

Figure (1.30): Second derivative spectra of mixtures contain (2-13) µg.mL^{-1} trifluoperazine hydrochloride in the presence of 2 µg.mL^{-1} olanzapine and carbamazepine at (A) (215-300) nm and (B) (251-285)nm.

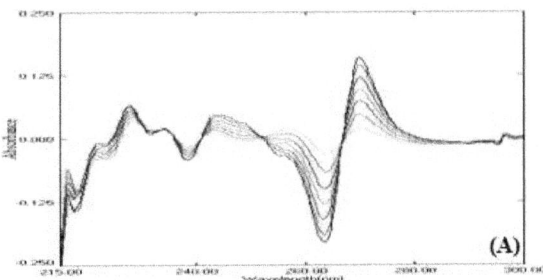

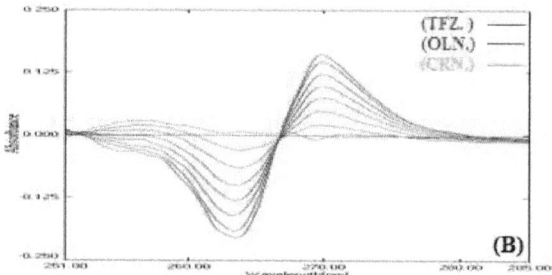

Figure (1.31): Second derivative spectra of mixtures contain (2-13) µg.mL^{-1} trifluoperazine hydrochloride in the presence of 6 µg.mL^{-1} olanzapine and carbamazepine at (A) (215-300) nm and (B) (251-285) nm.

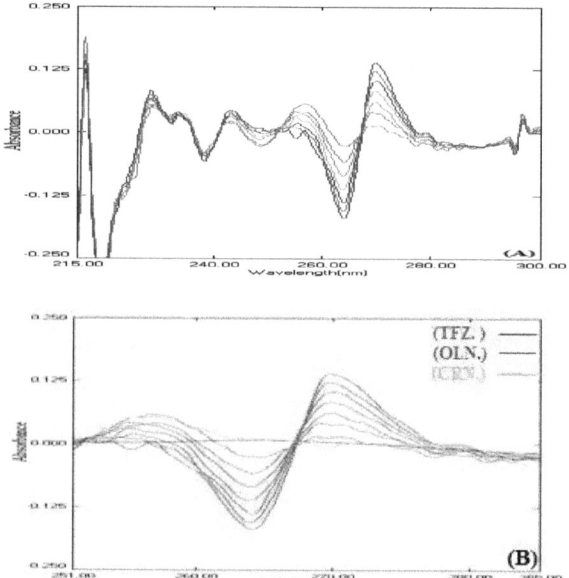

Figure (1.32): Second derivative spectra of mixtures contain (2-13) µg.mL^{-1} trifluoperazine hydrochloride in the presence of 10 µg.mL^{-1} olanzapine and carbamazepine at (A) (215-300) nm and (B) (251-285) nm.

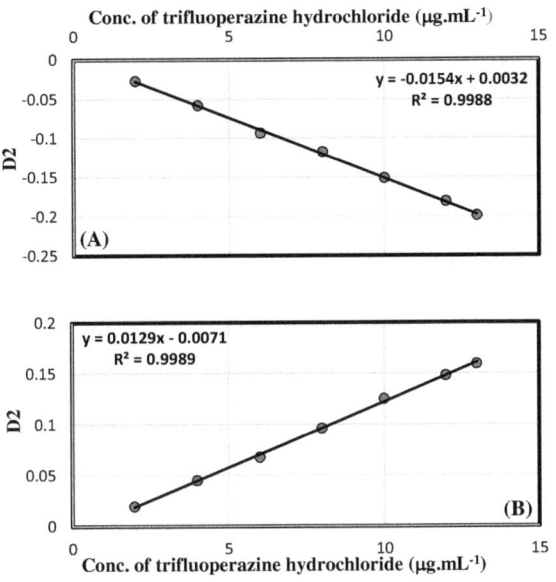

Figure (1.33): Calibration curves via second derivative spectra of (2-13) µg.mL^{-1} trifluoperazine hydrochloride for peak to baseline baseline at (A) 263.47 nm (B) 270.25 nm.

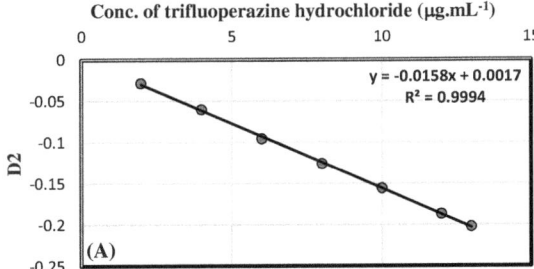

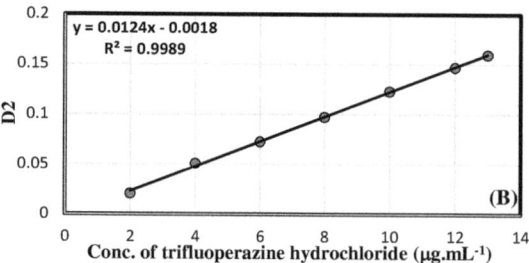

Figure (1.34): Calibration curves obtained via second derivative spectra of mixtures containing (2-13)µg.mL^{-1} trifluoperazine hydrochloride in the presence of 2 µg.mL^{-1} olanzapine and carbamazepine for peak-to-baseline at (A) 263.47 nm (B) 270.25 nm.

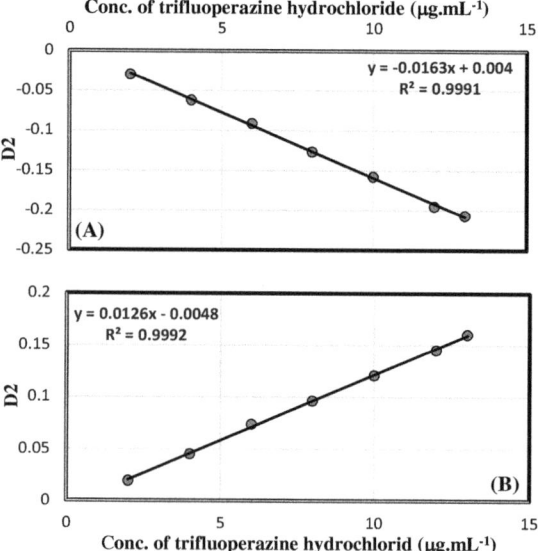

Figure (1.35): Calibration curves obtained via second derivative spectra of mixtures containing (2-13) µg.mL^{-1} trifluoperazine hydrochloride in the presence of 6 µg.mL^{-1} olanzapine and carbamazepine for peak-to-baseline at (A) 263.47 nm (B) 270.25 nm.

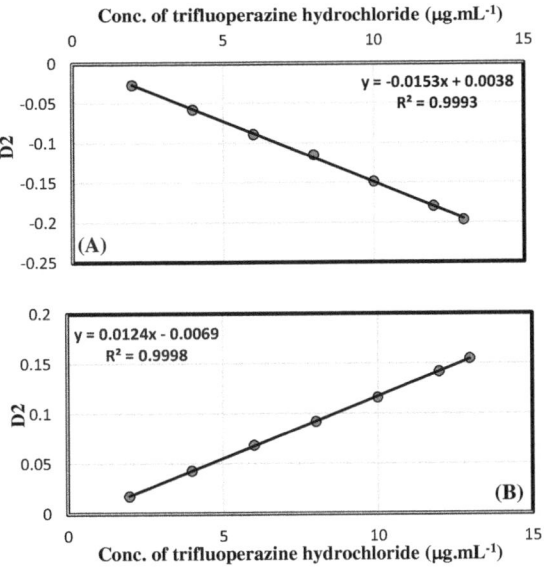

Figure (1.36): Calibration curves obtained via second derivative spectra of mixtures containing (2-13)μg.mL^{-1} trifluoperazine hydrochloride in the presence of 10 μg.mL^{-1} olanzapine and carbamazepine for peak-to-baseline at (A) 263.47 nm (B) 270.25 nm.

1.4.5 Calibration curves for carbamazepine

Under the mentioned optimum conditions for D1 and D2 modes in selected ranges of wavelengths calibration curves were obtained for the assay of carbamazepine via UV-spectrophotometric method.

The recorded spectra using D1 mode for a set of solutions containing (2- 14) μg.mL^{-1} carbamazepine in the presence of different concentrations of olanzapine and trifluoperazine hydrochloride (0, 2, 6 and 10) μg.mL^{-1} are shown in Figures (1.37)-(1.40). The procedure showed good results over the studied range of concentration depending on peak to baseline, for height to baseline at zero cross at 225.00 nm measurements, Figures (1.41)-(1.44) show plotted calibration curves..

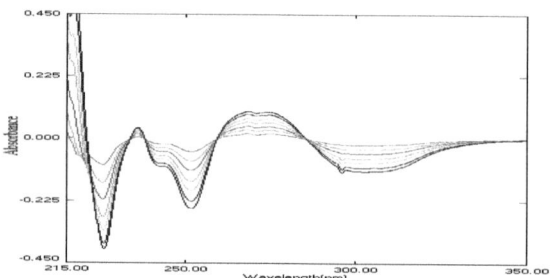

Figure (1.37): First derivative spectra of solutions containing (2-14) µg.mL^{-1} carbamazepine.

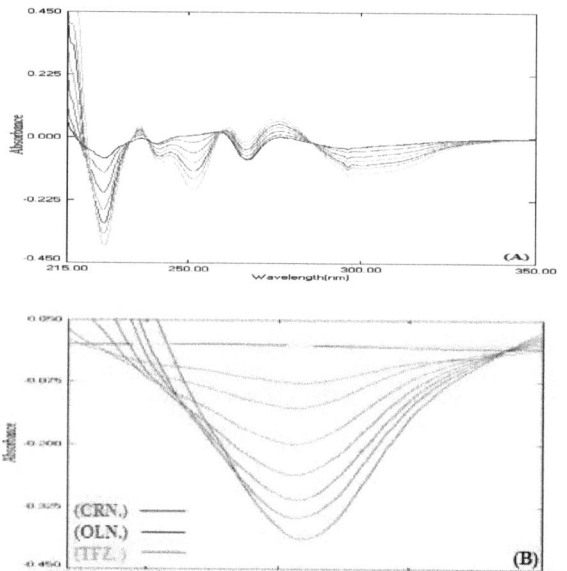

Figure (1.38): First derivative spectra of mixtures contain (2-14) µg.mL^{-1} carbamazepine in the presence of 2µg.mL^{-1} olanzapine and trifluoperazine hydrochloride (A) at (215- 350)nm, (B) at (217- 235)nm.

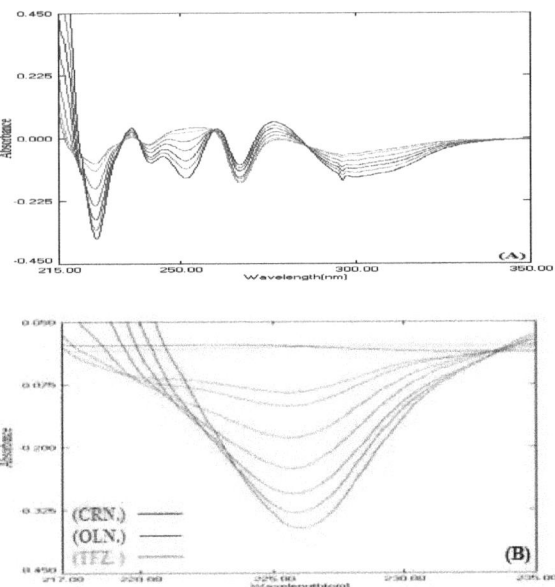

Figure (1.39): First derivative spectra of mixtures contain (2-14) µg.mL^{-1} Carbamazepine in the presence of 6 µg.mL^{-1} olanzapine and trifluoperazine hydrochloride (A) at (215- 350) nm, (B) at (217- 235) nm.

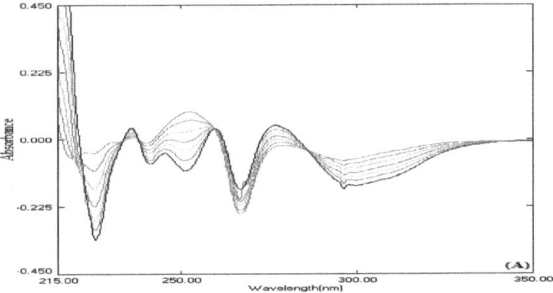

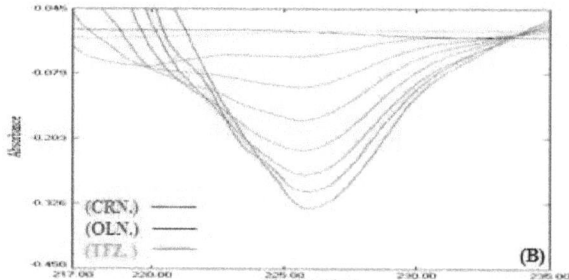

Figure (1.40): First derivative spectra of mixtures contain (2-14) µg.mL^{-1} carbamazepine in the presence of 10µg.mL^{-1} olanzapine and trifluoperazine hydrochloride (A) at (215- 350) nm, (B) at (217- 235)nm.

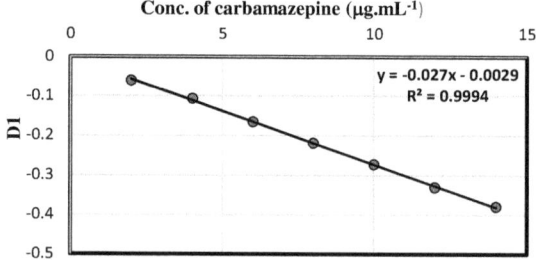

Figure (1.41): Calibration curves obtained via first derivative spectra of carbamazepine (2-14) µg.mL^{-1} for height to baseline at zero cross at 225.00 nm.

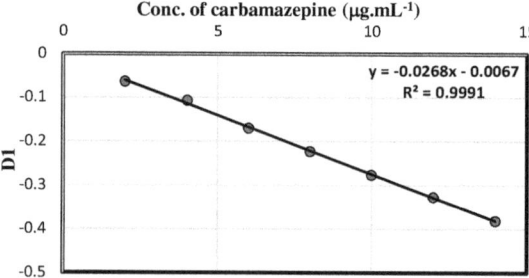

Figure (1.42): Calibration curve obtained via first derivative spectra of mixtures of carbamazepine (2-14) µg.mL^{-1} in the presence of 2 µg.mL^{-1} olanzapine and trifluoperazine hydrochloride for height to baseline at zero cross at 225.00 nm.

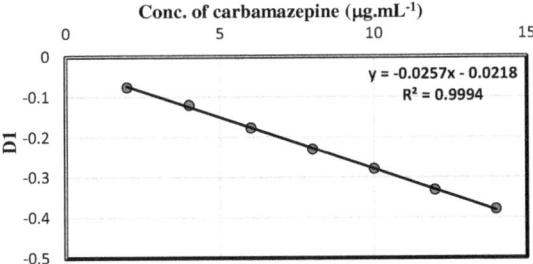

Figure (1.43): Calibration curve obtained via first derivative spectra of mixtures of carbamazepine (2-14) µg.mL^{-1} in the presence of 6 µg.mL^{-1} olanzapine and trifluoperazine hydrochloride for height to baseline at zero cross at 225.00 nm.

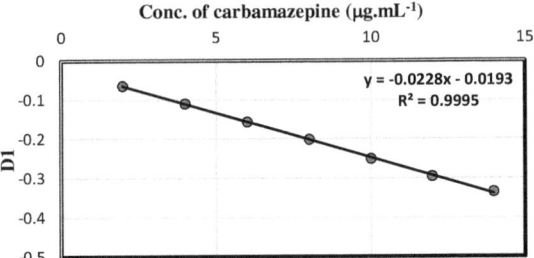

Figure (1.44): Calibration curves obtained via first derivative spectra of mixture of carbamazepine (2-14) µg.mL^{-1} in the presence of 10µg.mL^{-1} olanzapine and trifluoperazine hydrochloride for height to baseline at zero cross 225.00 nm.

The second derivative spectra obtained for the mixtures of (CRN.), (OLN.) and (TFZ.) show that height to baseline at zero cross at 231.27nm and peak to baseline at 238.69 nm, measurements could be used to quantify the exact concentration of (CRN.) in the presence of (OLN.) and (TFZ.). Figures (1.45)-(1.48) show the calculated second derivative spectra of different mixtures of the cited druges while, Figures (1.49)-(1.52) repesnt the calibration graphs for the same mixtures.

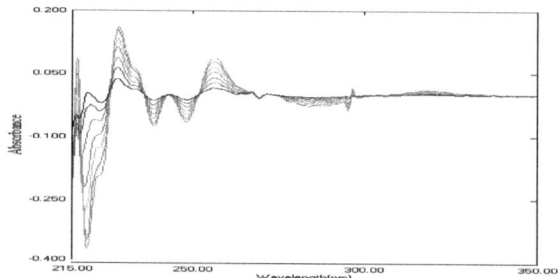

Figure (1.45): Second derivative spectra of solutions containing (2-14) µg.mL^{-1} carbamazepine.

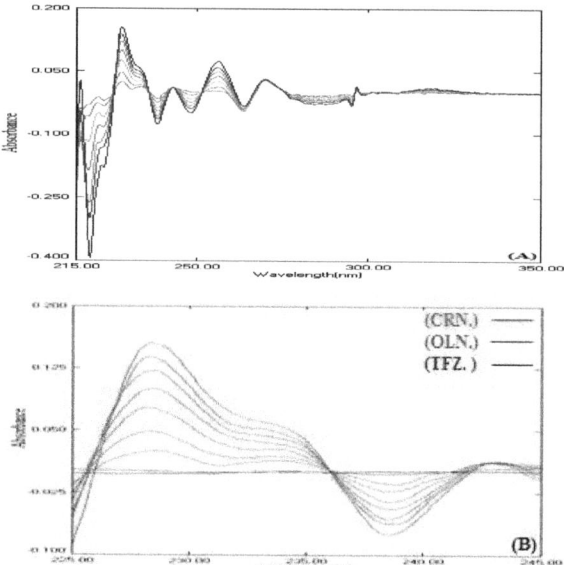

Figure (1.46): Second derivative spectra of mixtures contain (2-14) µg.mL^{-1} carbamazepine in the presence of 2µg.mL^{-1} olanzapine and trifluoperazine hydrochloride (A) at (215- 350)nm, (B) at (225- 245)nm.

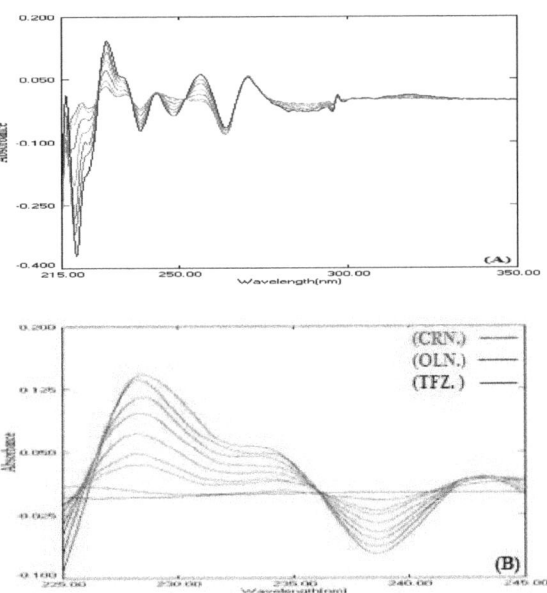

Figure (1.47): Second derivative spectra of mixtures contain (2-14) µg.mL^{-1} carbamazepine in the presence of 6 µg.mL^{-1} olanzapine and trifluoperazine hydrochloride (A) at (215- 350)nm, (B) at (225- 245)nm.

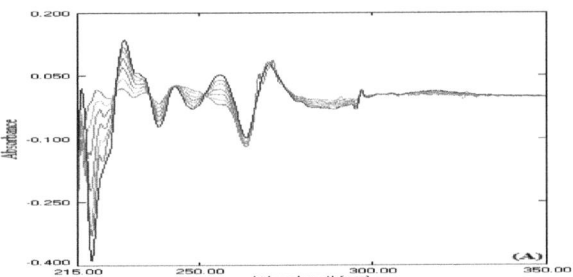

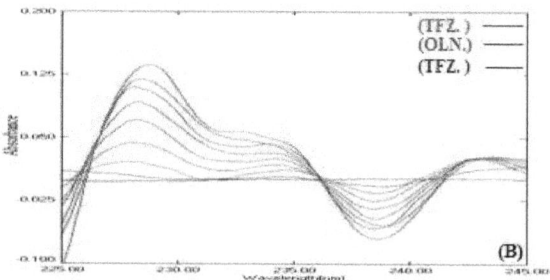

Figure (1.48): Second derivative spectra of mixtures contain (2-14) µg.mL⁻¹ carbamazepine in the presence of 10 µg.mL⁻¹ olanzapine and trifluoperazine hydrochloride (A) at (215- 350)nm, (B) at (225- 245)nm.

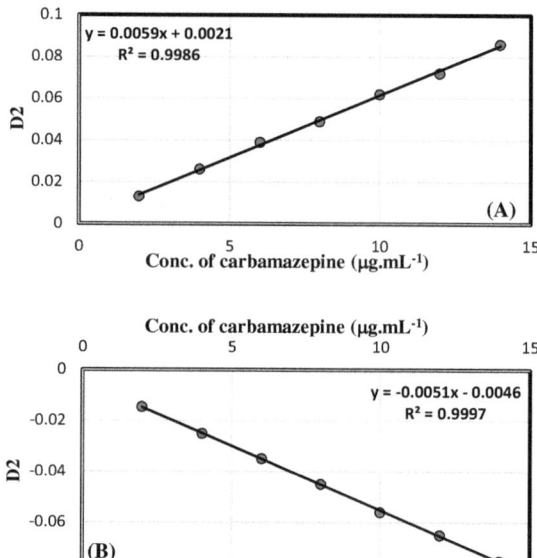

Figure (1.49): Calibration curves obtained via second derivative spectra of carbamazepine (2-14) µg.mL⁻¹ for (A) height to baseline at zero cross at 231.27 nm and (B) peak to baseline at 238.69 nm.

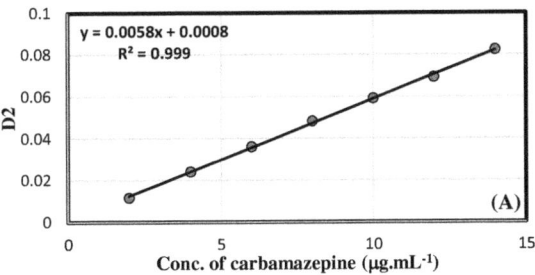

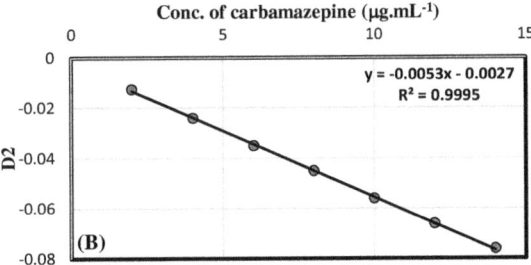

Figure (1.50): Calibration curves obtained via second derivative spectra of mixtures containing (2-14) µg.mL^{-1} carbamazepine in the presence of 2 µg.mL^{-1} olanzapine and trifluoperazine hydrochloride for (A) height to baseline at zero cross at 231.27 nm and (B) peak-to- baseline at 238.69 nm.

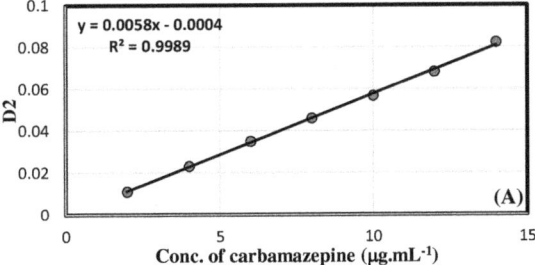

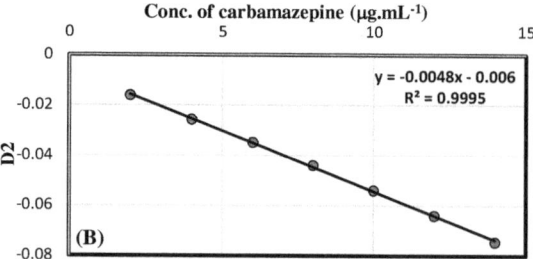

Figure (1.51): Calibration curves obtained via second derivative spectra of mixtures containing (2-14) µg.mL^{-1} carbamazepine in the presence of 6 µg.mL^{-1} olanzapine and trifluoperazine hydrochloride for (A) height to baseline at zero cross at 231.27 nm and (B) peak-to-baseline at 238.69 nm.

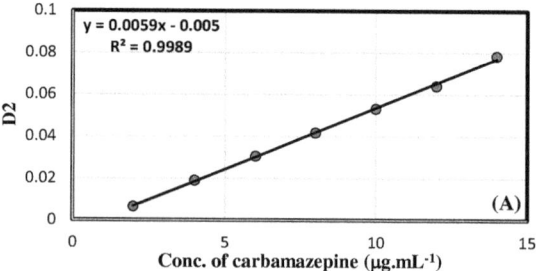

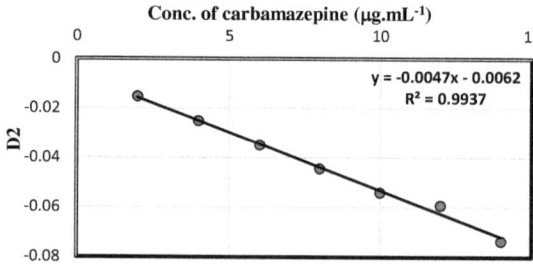

Figure (1.52): Calibration curves obtained via second derivative spectra of mixtures containing (2-14) µg.mL^{-1} carbamazepine in the presence of 10 µg.mL^{-1} olanzapine and trifluoperazine hydrochloride for (A) height to baseline at zero cross at 231.27 nm and (B) peak to baseline at 238.69 nm.

1.4.6 Precision and accuracy

The accuracy of the proposed methods was studied by carrying out three replicate analyses of two different amounts of each of the studied drug (within Beer's law), two drug solutions at studied concentration levels were analyzed each three times for both first and second order derivative spectrophotometric method, and the percentage relative error was calculated. Tables (1.6)-(1.8) shows all of the results.

The precision was determined in each case by calculating the coefficient of variation (C.V %) for the determinations at each of the studied concentration level.

Table (1.6): Evaluation of accuracy and precision for the determination of olanzapine by derivative technique.

Order of derivative	Mode of analysis	λ_{max} (nm)	Conc. of (OLN.) ($\mu g.mL^{-1}$)		Er%	C.V%
			Taken	Found*		
D1	Peak to baseline	291.47	5.000	4.979	-0.420	0.020
			10.000	9.992	-0.080	0.341
D2	Height to base line at zero cross	300.00	5.000	4.969	-0.620	0.403
			10.000	9.953	-0.470	0.250

*Average of three measurements.

Table (1.7): Evaluation of accuracy and precision for the determination of trifluoperazine hydrochloride by derivative technique.

Order of derivative	Mode of analysis	λ_{max} (nm)	Conc. of (TFZ.) ($\mu g.mL^{-1}$)		Er%	C.V%
			Taken	Found*		
D1	Peak to baseline	266.95	5.000	5.033	0.660	0.132
			10.000	10.045	0.450	0.392
D2	Peak to baseline	263.47	5.000	5.029	0.580	0.050
			10.000	10.067	0.670	0.145
		270.25	5.000	5.041	0.820	0.545
			10.000	10.055	0.550	0.234

*Average of three measurements.

Table (1.8): Evaluation of accuracy and precision for the determination of carbamazepine by derivative technique.

Order of derivative	Mode of analysis	λmax (nm)	Conc. of (CRN.) (µg.mL^{-1})		Er%	C.V%
			Taken	Found*		
D1	Height to base line at zero cross	225.00	5.000	4.961	-0.780	0.462
			10.000	9.984	-0.160	0.325
D2	Height to base line at zero cross	231.27	5.000	4.977	-0.460	0.277
			10.000	9.982	-0.180	0.097
	Peak to baseline	238.69	5.000	4.970	-0.600	0.302
			10.000	9.993	-0.070	0.680

*Average of three measurements.

1.4.7 Interference study

Specificity of the suggested method was tested by analyzing each drug in the presence of the other and in the presence of excipients that may be used in the dosage forms and the percentage recovery was calculated. This study was performed by the addition of known amounts of excipients to mixture solutions of the three examined drugs. First and second derivative techniques were used at the selected wavelength for the concentrations of drugs measurement. The high recovery showed that no interference were found using first and second derivative mode for the determination of (OLN.), (TFZ.) and (CRN.) in their mixture even in the presence of the added excipients, these results are listed in Table (1.9).

Table (1.9): Recovery values for mixtures of olanzapine, trifluoperazine hydrochloride and carbamazepine in the presence of 500 μg.mL⁻¹ of excipients.

Drug	Excipients	D1		D2	
		Found (µg.mL⁻¹)	Recovery %	Found (µg.mL⁻¹)	Recovery %
(OLN.)	Lactose	9.886	98.860	9.959	99.590
	Glucose	9.974	99.740	9.993	99.930
	Sucrose	9.892	98.920	9.937	99.370
	Starch	9.792	97.920	9.851	98.510
	Magnesium Stearate	9.977	99.770	9.993	99.930
(TFZ.)	Lactose	9.981	99.810	9.964	99.640
	Glucose	9.979	99.790	9.985	99.850
	Sucrose	10.071	100.710	10.015	100.150
	Starch	9.992	99.920	9.859	98.590
	Magnesium Stearate	10.035	100.350	10.062	100.620
(CRN.)	Lactose	9.958	99.580	9.981	99.810
	Glucose	9.993	99.930	9.974	99.740
	Sucrose	9.869	98.690	9.938	99.380
	Starch	10.057	100.570	10.032	100.320
	Magnesium Stearate	9.947	99.470	9.985	99.850

- D1 and D2 for 10 µg.mL⁻¹ (OLN.) peak to baseline and height to baseline at zero cross, at 291.47 nm and 300.00 nm respectively.
- D1 and D2 for 10 µg.mL⁻¹ (TFZ.) peak to baseline at 266.95 nm and 263.47 nm respectively.
- D1 and D2 for 10 µg.mL⁻¹ (CRN.) height to baseline at zero cross at 225.00 nm and 231.27 nm respectively.

1.4.8 Application to pharmaceutical preparation

In order to evaluate the efficiency of the derivative technique in the determination of (OLN.), (TFZ.) and (CRN.) drugs in pharmaceutical formulations, D1 and D2 procedures were applied. Good recovery % and C.V% values indicated the suitability of these methods for routine analysis of (OLN.), (TFZ.) and (CRN.). The summary of the results is depicted in Table (1.10).

Table (1.10): Statistical validation data for quantitative assessment of commercial tablet formulations for (OLN.), (TFZ.) and (CRN.)

Drug	Tablet Sample	Weight Labeled (mg/tablet)	Weight Found* (mg/tablet)			Mean (mg/tablet)	Recovery %	C.V%
			D1*					
(OLN.)	Zyprexa- 5 mg	5.000	5.107	5.042	5.089	5.079	101.580	0.660
	Zyprexa- 10 mg	10.000	10.091	10.105	10.100	10.098	100.980	0.070
	Olan- 5 mg	5.000	4.630	4.801	4.817	4.749	94.980	2.182
(TFZ.)	Iralzine - 5 mg	5.000	5.010	5.016	5.021	5.015	100.300	0.109
	Espazine - 5mg	5.000	5.070	5.141	5.190	5.133	102.660	1.175
	Stellasil - 1 mg	1.000	1.052	1.061	1.011	1.041	104.100	2.560
(CRN.)	Carbasam - 200 mg	200.000	200.150	200.133	200.807	200.363	100.181	0.191
	Tegretol - 200mg	200.000	203.077	208.100	210.620	207.265	103.632	1.852
			D2*					
(OLN.)	Zyprexa- 5 mg	5.000	5.089	5.144	5.109	5.114	102.280	0.544
	Zyprexa- 10 mg	10.000	10.217	10.162	10.300	10.226	10.260	0.679
	Olan- 5 mg	5.000	4.815	4.670	4.601	4.695	93.900	2.321
(TFZ.)	Iralzine - 5 mg	5.000	5.150	5.194	5.204	5.182	103.640	0.554
	Espazine - 5mg	5.000	5.203	5.219	5.145	5.189	103.780	0.750
	Stellasil - 1 mg	1.000	1.100	1.093	1.027	1.060	106.000	3.490
(CRN.)	Carbasam - 200 mg	200.000	203.510	203.822	205.890	204.890	102.445	0.631
	Tegretol - 200mg	200.000	207.050	211.006	209.722	209.259	104.629	0.964
			D2**					
(TFZ.)	Iralzine - 5 mg	5.000	5.027	5.050	5.044	5.040	100.800	0.236
	Espazine - 5mg	5.000	5.110	5.017	5.069	5.065	101.300	0.920
	Stellasil - 1 mg	1.000	1.030	1.032	1.043	1.035	103.500	0.676
(CRN.)	Carbasam - 200 mg	200.000	202.090	205.024	205.470	204.194	102.097	0.899
	Tegretol - 200mg	200.000	206.160	209.440	207.002	207.534	103.767	0.820

*D1 and *D2 for (OLN.) peak to baseline and height to baseline at zero cross, at 291.47 nm and 300.00 nm respectively.

*D1, *D2 and **D2 for (TFZ.) peak to baseline at 266.95 nm, 263.47 nm, 270.25nm respectively.

*D1, *D2 and **D2 for (CRN.) height to baseline at zero cross at 225.00 nm and 231.27nm, and peak to baseline at 238.69 nm. respectively.

References

1. Patel V. M.; Patel J. A.; Havele S. S. and Dhaneshwar S. R. *"First and Second Derivative Spectrophotometric methods for Determination of Olanzapine in Pharmaceutical"* International J. of ChemTech Research, **2**(1), (2010), p: (756-761).

2. Karpinska J. *"Basic Principles and Analytical Application of Derivative Spectrophotometry"* Macro to nano spectroscopy, book edited by Jamal Uddin, (2012), p: (253-256).

3. Talsky G. *"Derivative Spectrophotometry"* 1st edition, VCH, Weinheim, (1994).

4. Owen A. J. *"Uses of Derivative Spectroscopy"* UV-Visible Spectroscopy, Agilent Technologies, (1995).

5. Tony O. *"Obtaining derivative spectra- Fundamentals of modern UV-visible spectroscopy"* Agilent Technologies, **8**, (2000).

6. Al-saidi H. K. *"Quantitative determination of some antibiotic drugs of penicillin and cephalosporin groups through a new selective electrodes and derivative spectrophotometry"*, A thesis Submitted to the College of Sci. Al-mustansiria University, (2006), p: (20).

7. El-osayed A. Y., and El-Salem N. A. *"Recent Development of Derivative Spectrophotometry and their Analytical Applications"*, J. of Anal. Scince, **21**, (2005) p: (595-564).

8. Eskandari H., Saghseloo A. G. and Chamjangali M. A., *"First- and Second-Derivative Spectrophotometry for Simultaneous Determination of Copper and Cobalt by 1-(2-Pyridylazo)-2-naphthol in Tween 80 Micellar Solutions"* Turk J. Chem., **30**, (2006), p: (49 -63).

9. Stanisz B.; Paszun S. and Leśniak M. *"Validation of UV Derivative Spectrophotometric Method for Determination of Benazepril Hydrochlodide in Tablets and Evaluation of Its Stability"*, Acta Poloniae Pharmaceutica-Drug Research, **66**(4), (2009), p: (343-349).

10. Ojeda C. B. and Rojas F. S. *"Recent developments in derivative ultraviolet-visible absorption spectrophoto-metry"*, Analytica Chimica Acta, **518**, (2004), p: (1-24).

11. Dikran S. B., and Mohammed J. M., *"First and second Derivative Spectrophotometry for individual and simultaneous determination of Amoxicillin and Cephalexin"*, national J. of Chemistry, **34**, (2009), p: (260-269).

12. Koranya M. A.; Abdinea H. H.; Ragaba M. A. A. and Aborassa S. I., *"Application of Derivative Spectrophotometry under orthogonal polynomial at unequal intervals: Determination of metronidazole and nystatin in their pharmaceutical mixture"*, Spectrochimica Acta Part A: Molecular and Biomolecular Spectroscopy, **143**, (2015),p: (281–287).

13. Vichare V.; Suryawanshi P.; Bhosale J. and Dhole S., *"Simultaneous Estimation of Metformin HCl and Pioglitazone HCl by Second Order Derivative UV-Visible Spectrophotometric Method in Tablet Formulation"*, Asian J. Pharm. Tech., **4** (3), (2014), p: (157-162).

14. Wahab S. I. and Imran T. *"First Order Derivative Spectrophotometric Method Develop and Validate for Estimation of Bifonazole in Bulk Drug and Pharmaceutical Formulation"*, International J. of Pharm. Research & Allied Sciences, **2**, (2013), p: (60-64).

15. Baghdadya Y. Z.; Al-Ghobashyb M. A.; Abdel-Aleemb A. E. and Weshahya S. A. *"Spectrophotometric and TLC-densitometric methods for the simultaneous determination of Ezetimibe and Atorvastatin calcium"* J. of Advanced Research, **4**(1), (2013), p: (51-59).

16. Attimarad M.; Al-Dhubiab B. E.; Alhaider I. A.; Nair A. B.; Harsha N. S. and Ahmed K. M. *"Simultaneous determination of moxifloxacin and cefixime by first and ratio first derivative ultraviolet*

spectrophotometry" Attimarad et al. Chem. Central J., **6**(105), (2012), p: (1-7).

17. Lakiss H.; Ilie M.; Baconi D. L. and Bălălău D. *"Derivative UV Spectrophotometry Used for The Assay of Diazepam from Human Blood Plasma"*, Farmacia, **60**(4), (2012), p: (565-570).

18. Dhanawade P. P. and Kane R. N. *"Derivative Spectro-photometric Method for Estimation of Irbesartan in Bulk Drug and Dosage form"* International J. of Research in Pharm. and Biomedical Sci., **3**(3), (2012), p: (1300-1305).

19. Khamar J. C. and Pate S. A. *"First Derivative Spectro-photometric Method for The Simultaneous Estimation of Rifampicin and Piperine in Their Combined Capsule Dosage Form"* Asian J. of Pharmacy and Life Sci., **2**(1), (2012), p: (50-55).

20. Saraan S. M. D.; Sinaga M. and Muchlisyam A. *" Development Method for Determination of Ternary Mixture of Paracetamol, Ibuprofen and Caffeine in Tablet Dosage Form Using Zero-crossing Derivative Spectrophotometric"* Int. J. Pharm Tech Res., **7**(2), (2015), p: (349-353).

21. Aydolmu Z. G.; YJlmaz E. M.; YJlmaz S. and AkpJnar S. *"Development of Simultaneous Derivative Spectrophotometric and HPLC Methods for Determination of 17-Beta-Estradiol and Drospirenone in Combined Dosage Form"* International Scholarly Research Notices, **2015**, (2015), p: (1-7).

22. El-Zinati A. M. and Abdel-Latif M. S. *"Simultaneous Determination of Paracetamol and Tramadol in Pharmaceutical Tablets by Derivative UV-Vis Absorption Spectrophotometry"* The Open Analytical Chemistry Journal, **8**, (2015), p: (1-6).

23. Pradhan K. P.; Raiyani R.; Shah S. R.; Grishma H. P. and Umesh U. *"Second derivative spectrophotometric method development and*

validation for simultaneous estimation of gatifloxacin and prednisolone acetate in their combined dosage form" The Pharma Innovation Journal, **3**(11), (2015), p: (6-10).

24. Souri E.; Rahimi A.; Ravari N. S. and Tehrani M. B. *"Development of a Rapid Derivative Spectrophotometric Method for Simultaneous Determination of Acetaminophen, Diphenhydramine and Pseudoephedrine in Tablets"* Iranian Journal of Pharmaceutical Research, 14 (2), (2015), p: (435-442).

25. Hoang V. D.; HaLy D. T.; HuuTho N. and Nguyen H. M. T. *" UV Spectrophotometric Simultaneous Determination of Paracetamol and Ibuprofen in Combined Tablets by Derivative and Wavelet Transforms"* The Scientific World Journal, **2014**, (2014), p: (1-13).

26. Ahluwalia V. K. and Raghav S. *"Comprehensive Experimental Chemistry"* New Age International Limited, Puplishers, printed in India, (1997), p: (13-14).

List of contents

Subject number	Title	Page number
1.	Spectrophotometric simultaneous determination of olanzapine trifluoperazine hydrochloride and carbamazepine via derivative spectrophotometry	1
1.1	Introduction	1
1.2	Estimation of some drugs in dosage compounds and human fluids by derivative mode	3
1.3	Experimental	5
1.3.1	Instruments	5
1.3.2	Chemical compounds	5
1.3.3	The drugs and pharmaceutical preparations	6
1.3.4	Preparation of solutions	6
1.3.5	Preparation of standard drugs solutions	7
1.3.5.1	Olanzapine stock solution (1000 µg.mL^{-1})	7
1.3.5.2	trifluoperazine hydrochloride stock solution (1000 µg.mL^{-1})	7
1.3.5.3	carbamazepine stock solution (1000 µg.mL^{-1})	7
1.3.6	Solutions for the analysis of drugs in pharmaceutical preparations	8
1.3.6.1	Olanzapine	8
1.3.6.2	trifluoperazine hydrochloride	8
1.3.6.3	Carbamazepine	9
1.3.7	Recommended procedures	9
1.3.7.1	Assay procedure for the determination of olanzapine, trifluoperazine hydrochloride and carbamazepine	9
1.3.7.2	Assay procedure for the determination of the drugs mixtures	10
1.4	Results and discussion	11
1.4.1	Absorption spectra at zero order mode	11
1.4.2	First and second derivative modes	11
1.4.3	Calibration curves for olanzapine	13
1.4.4	Calibration curves for trifluoperazine hydrochloride	22
1.4.5	Calibration curves for carbamazepine	31
1.4.6	Precision and accuracy	41
1.4.7	Interference study	42
1.4.8	Application to pharmaceutical preparation	43
	References	45

I want morebooks!

Buy your books fast and straightforward online - at one of world's fastest growing online book stores! Environmentally sound due to Print-on-Demand technologies.

Buy your books online at
www.morebooks.shop

Kaufen Sie Ihre Bücher schnell und unkompliziert online – auf einer der am schnellsten wachsenden Buchhandelsplattformen weltweit! Dank Print-On-Demand umwelt- und ressourcenschonend produziert.

Bücher schneller online kaufen
www.morebooks.shop

KS OmniScriptum Publishing
Brivibas gatve 197
LV-1039 Riga, Latvia
Telefax: +371 686 204 55

info@omniscriptum.com
www.omniscriptum.com

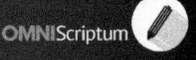

Milton Keynes UK
Ingram Content Group UK Ltd.
UKHW040705050124
435493UK00001B/236